SPACE BUSTERS
SPACE DRAMAS

Chris Woodford

Chrysalis Children's Books

LOOK FOR THE RINGS OF SATURN

Look out for boxes like this with the planet Saturn in the corner. They contain extra information and amazing space-buster facts and figures.

Produced by
Monkey Puzzle Media Ltd,
Gissing's Farm, Fressingfield,
Suffolk IP21 5SH, UK

First published in Great Britain
in 2002 by
 Chrysalis Children's Books
64 Brewery Road,
London N7 9NT

Copyright © Chrysalis Books PLC
2002

Paperback edition first published
in 2003

A Belitha Book

Designer: Tessa Barwick
Editor: Sarah Doughty
Consultant: Stuart Atkinson

ISBN 1 84138 364 3 (hb)
ISBN 1 84138 769 X (pb)

British Library Cataloguing in
Publication Data for this book is
available from the British Library.

Printed in Hong Kong
10 9 8 7 6 5 4 3 2 1 (hb)
10 9 8 7 6 5 4 3 2 1 (pb)

Acknowledgements
We wish to thank the following
individuals and organizations for
their help and assistance and for
supplying material in their
collections: Camera Press back
cover top (Lucasfilm Ltd), 5 bottom
(Lucasfilm Ltd), 6 left (Luc Hieulle),
16 top (W Drommer), 16 bottom
(Lucasfilm Ltd), 21 bottom (Jerry
Watson), 23 bottom, 27 (Tom
Keller); Corbis back cover bottom
left (Forrest J Ackerman Collection),
4 (Bettmann), 7 (Forrest J Ackerman
Collection), 13 bottom, 17
(Bettmann), 22 (George Hall),
25 left (Richard Cummins), 26
(L Clarke); Kobal Collection 18
(Peter Mayhew); MPM Images 3,
10, 11 centre, 11 bottom, 14, 28,
29; NASA 23 top; Science Photo
Library front cover (Peter Menzel),
back cover bottom right (Stuart
Painter), 3 (Peter Menzel), 5 top
(Stuart Painter), 8 left (Peter
Menzel), 9 (Magrath/Folsom),
11 top (Peter Menzel), 12 (Julian
Baum), 13 top (David A Hardy),
15 (A Gragera/Latin Stock), 19 top
(Biophoto Associates), 24 (Victor
Habbick Visions), 25 right (NASA);
Topham Picturepoint 1, 6 right, 8
right, 19 bottom, 20, 21 top, 30.

▼ The *Ranger 7* spacecraft sent thousands of photographs back to Earth before crashing into the Moon in 1964.

▶ Astronauts have to carry out many dangerous spacewalks to build the new International Space Station.

CONTENTS

THE DRAMA OF SPACE

Have you ever felt scared going into a dark room? Imagine then how astronauts must feel when they set off in a rocket into space. The world of stars and planets is dark and distant and it is full of danger and excitement.

A rocket launches astronauts into space with enormous power and speed. But a rocket is a complicated machine and many different things can go wrong. Reaching space safely is only the start of the astronauts' mission (journey into space).

◀ Many films, such as *Mission to Mars,* have been made about space and what it might be like.

Astronauts often have to do a number of dangerous jobs in space. They are a long way from Earth if anything goes wrong, and sometimes it can be difficult to get them back to Earth safely. Fortunately, most of the astronauts who travel into space live to tell their dramatic tales!

DREAMY DRAMAS

In the seventeenth century, a German astronomer called Johannes Kepler tried to imagine what travelling into space would be like. He thought people might go to the Moon on a bridge made out of a shadow and find it full of scary creatures like huge snakes with wings. Fortunately, nothing like this was found when astronauts reached the Moon in 1969!

◀ Space can seem a scary place when you're millions of kilometres from home.

COUNTDOWN TO DANGER

▼ Blast-off is a very carefully controlled explosion.

When a rocket blasts off from the launchpad (its firing base) no one can tell if it will shoot safely into space or explode in a huge fireball. For an astronaut, blasting off on a rocket is a bit like sitting on a giant firework packed with explosives!

Sometimes launches end in disaster. In 1998, a Titan 4A rocket blew up just 42 seconds after its launch from Cape Canaveral in Florida, USA. The explosion destroyed a $1 billion satellite that was on board.

THE CHALLENGER DISASTER

The most shocking space disaster so far happened on 28 January, 1986. Just 73 seconds after blast-off, the space shuttle *Challenger* exploded in mid-air because of a fault in one of its rocket boosters. The world was stunned by the terrible accident in which all seven of the astronauts on board were killed.

▲ A European Ariane rocket blazes
a trail as it lifts off into the sky.

Two years earlier, a faulty
computer program caused the
European *Ariane 5* rocket to
go off course. The rocket was
worth $500 million and had
taken ten years to develop. It
had to be destroyed by people
on the ground just 40 seconds
after blast-off.

▶ This satellite dish is used to control
Ariane rockets from Earth.

STEPPING INTO SPACE

Inside a rocket, astronauts are protected from any dangers. But sometimes they have to go outside the spacecraft to explore space or to make repairs.

Stepping outside a spacecraft is called a 'spacewalk'. The astronauts look as though they are floating in space, but they are usually tied to their craft by long support lines called umbilicals. Astronauts wear space suits to protect their bodies from the incredible heat and cold, and from tiny meteorites (flying space rocks) that may crash into them. Large helmets like goldfish bowls provide them with oxygen. Gold visors stop the Sun's dangerous rays from hurting their eyes.

In 1965, the Russian astronaut Alexei Leonov became the first person to walk in space. The next year, American astronaut Edwin 'Buzz' Aldrin made a spacewalk lasting two hours, 20 minutes.

MAKING SPACE SAFER

The worst thing that could happen during a spacewalk would be for an astronaut's umbilical to break. The astronaut might drift off into space, never to be seen again. Fortunately, NASA (the US space organization) has now developed a special kind of jetpack called SAFER. If an accident happens, the astronaut can use this to fly back safely to the spacecraft.

▼ Preparing for a disaster is part of every astronaut's training. Here, two astronauts practise the SAFER test during a spacewalk.

Lost in Space

Spacecraft are sometimes controlled by the astronauts on board and sometimes by mission control (the control centre on Earth). When contact between mission control and a spacecraft is lost, the results can be dramatic.

Space probes are spacecraft with no astronauts on board. They are sent out to explore the Universe. In 1999, NASA lost a $125 million probe called the *Mars Climate Orbiter* because of a simple mistake.

◄ Expensive spacecraft sometimes disappear into space, never to be seen again.

The NASA scientists had been using metric measurements (metres and centimetres), while the satellite's builders had used standard American measurements (feet and inches). As a result, the satellite was steered too close to Mars and burned up in the Martian atmosphere!

▶ The *Hipparcos* satellite was almost lost when it was launched into the wrong orbit in 1989.

STRANDED ON THE MOON?

When the Apollo astronauts set off for the Moon in 1969, some people worried they might never return. America's president, Richard Nixon, had prepared a speech to read out to the world if Neil Armstrong and his colleagues became stuck on the Moon. Fortunately, the mission was a success and the astronauts returned home safely.

◀ Scientists monitor, or check, spacecraft from their screens at mission control.

BIG BANGS

Most of the space dramas we hear about happen when astronauts set out to explore the Universe. But some very dramatic things happen naturally in space. These include the 'big bangs' caused by meteors and solar flares.

Meteors are lumps of rock that travel through space. When a meteor hits Earth, it can make a massive pit, or crater. One of the best known meteor craters on Earth is Barringer's Crater in Arizona. It is thought to have been made at least 50 000 years ago.

Some scientists believe the dinosaurs became extinct on Earth when a giant meteor crashed into our planet 65 million years ago. The chances of this happening again are very small, because the Universe is such a big place.

▼ **Barringer's Crater in Arizona, USA is 800m wide and 200m deep.**

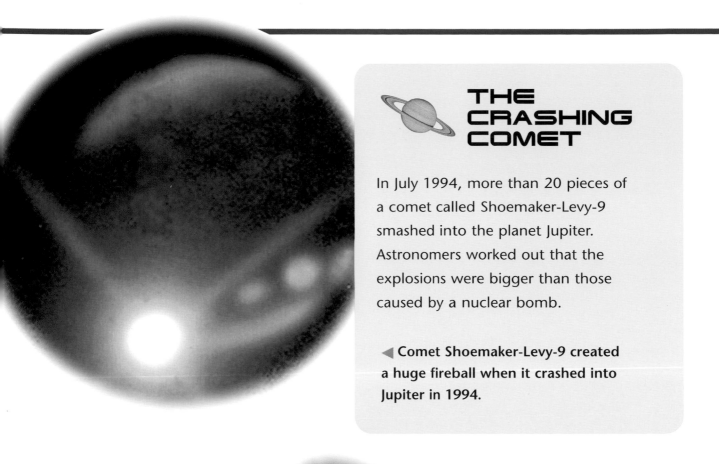

THE CRASHING COMET

In July 1994, more than 20 pieces of a comet called Shoemaker-Levy-9 smashed into the planet Jupiter. Astronomers worked out that the explosions were bigger than those caused by a nuclear bomb.

◀ Comet Shoemaker-Levy-9 created a huge fireball when it crashed into Jupiter in 1994.

▼ Solar flares are massive bursts of energy from the surface of the Sun.

The biggest bangs in the Solar System are caused by solar flares. These are tremendous explosions taking place every so often from the surface of the Sun. Each one has the same power as millions of large nuclear bombs exploding all at once!

DANGEROUS DOCKINGS

▼ Russian and American astronauts met up in space for the first time in July 1975, when their Apollo and Soyuz spacecrafts docked.

Spacecraft need to join together, or dock, with the space station, to allow people to get on and off the space station. Docking is difficult and dangerous, and sometimes things go wrong.

The very first docking of two spacecraft happened in 1966. The rocket, *Gemini 8*, was piloted by astronaut Neil Armstrong. It was supposed to dock with an unmanned craft called the GATV (Gemini Agena target vehicle). It took six hours for the craft to meet up, and then disaster struck.

▼ This photograph of the GATV is what Neil Armstrong saw from *Gemini 8* in 1966, just before the two craft docked.

A fuel leak made the two docked spacecraft spin round and round. Armstrong decided to separate them, but this only made things worse. Soon, his spacecraft was spinning so fast (once a second) that Armstrong and his co-pilot could no longer see properly. Armstrong had no choice but to end the mission. *Gemini 8* made an emergency splashdown in the Pacific Ocean and the astronauts were recovered safely.

 A HIT FOR MIR

Russia's space station, *Mir*, provided lots of drama during its 15 years in space. One of the biggest dramas happened in 1997, when an unmanned cargo ship weighing over seven tonnes crashed into *Mir* during a docking exercise. The Russian cosmonauts and US astronauts on board had just half an hour to fix the damage. Later, they had to carry out a dangerous spacewalk to turn *Mir*'s solar panels towards the Sun so the craft could regain power.

Close Encounters

▲ Halley's Comet photographed by *Giotto*.

Space probes are often sent on missions to find out more about comets or asteroids. Comets are sometimes called dirty snowballs, because they are made up of ice and dust. They have a central section called the nucleus and a long tail of dust and gas.

Scientists found out much more about comets by sending a space probe called *Giotto* to meet the most famous comet of all, Halley's Comet. This close encounter happened 93 million kilometres from Earth in March 1986. *Giotto* passed within 605 kilometres of the comet's nucleus (the leading 'snowball') and was knocked about by the dusty wind from the comet as it passed by.

◄ *Giotto* was travelling faster than a bullet when it passed Halley's Comet in 1986.

16

The NEAR probe prepares for its historic landing on the Eros asteroid in 2001.

NEAR AND FAR

On 12 February 2001, a small probe called *Near Earth Asteroid Rendezvous* (NEAR) *Shoemaker* managed to land on top of an asteroid called Eros after circling around it for about a year. This dramatic encounter happened 193 million kilometres from Earth.

Asteroids are small planets that circle the Sun in a broad band between the planets Mars and Jupiter. In 1991, the space probe *Galileo* made the first close-up study of an asteroid called Gaspra during its voyage to study Jupiter. *Galileo* and Gaspra flew past one another at 29 000 km/h. (That's over ten times faster than the top speed of a jet fighter!)

DARING REPAIRS

Spacecraft are incredibly complicated machines. When things go wrong with them, astronauts may need to risk their lives making repairs. But even when the repairs have been made, other things may go wrong!

HUBBLE TROUBLE

When the Hubble Space Telescope (HST) was launched in 1990, astronomers were shocked to find that all the pictures it sent to Earth were blurred. The telescope's mirror had been made slightly too flat. It took five walks in space in 1993 to repair the Hubble. But it now sends back dramatic pictures of the Universe... in perfect focus!

◄ An astronaut treads carefully as he carries out repairs on the Hubble Telescope in the bay of the space shuttle.

▲ Even space repairmen need their tools. This is a battery-powered screwdriver used by astronauts.

NASA's first space station, *Skylab 1*, was damaged when it took off in May 1973. As it flew into orbit, an important shield was ripped away from the craft. This had been designed to protect *Skylab* from the heat of the Sun and from meteorites. When the shield broke away, it took one of *Skylab's* solar panels (long flat panels that use the Sun's energy to make power) with it and pushed the other one out of position. The craft was left without power.

Three astronauts took off in another rocket to make repairs. One of them leant out of an open hatch and used a large hook to pull *Skylab's* solar panel back into position. Next, the astronauts put up a giant umbrella over the outside of *Skylab* to stop it being damaged by the Sun.

◄ The first repair of a satellite in space took place in 1984.

RISKY RESCUES

When accidents happen in space, the most important thing is to get the astronauts safely back to Earth. All astronauts prepare for emergencies as part of their training. Even so, space rescues are always difficult and dangerous.

American astronaut Virgil 'Gus' Grissom almost lost his life in July 1961 when his Mercury capsule splashed down in the Atlantic Ocean. The hatch flew open too soon, causing the capsule to fill with water and then sink. Grissom almost drowned before being rescued by a helicopter.

▲ A US marine helicopter tries to rescue Gus Grissom from the Atlantic Ocean.

▶ Astronauts train in giant swimming pools on Earth. Here, they are preparing for an emergency on the International Space Station.

Sometimes accidents happen too quickly for astronauts to be rescued. In 1971, the Russian *Soyuz 11* spacecraft was on its way back to Earth when something went wrong. An air valve blew open, causing all the air in the capsule to be sucked out. The three cosmonauts on board were killed by the lack of oxygen as their craft sped back to Earth.

 ## APOLLO RETURNS

Disaster struck the *Apollo 13* mission to the Moon in April 1970. Halfway to the Moon, a tank of oxygen exploded. *Apollo 13* lost power and the astronauts had to be rescued by mission control at Houston, Texas, USA. Millions of people around the world watched the rescue on television and prayed for the astronauts' safe return. The rescue was a success and *Apollo 13* splashed down safely in the South Pacific Ocean four days later.

▶ **Astronaut John Swigert is rescued from the ocean by a helicopter, following the *Apollo 13* splashdown.**

UNHAPPY LANDINGS

Spacecraft that take off from Earth do not always come back again. Some stay in space forever. Others burn up in Earth's atmosphere on the way back. And some come crashing back to Earth when they are least expected!

Many people were worried when NASA decided to bring its *Skylab 1* space station back to Earth in 1979. NASA was unable to tell where the station would crash land. Fortunately, most of the craft burned up as it passed back into Earth's atmosphere, but some did come back to Earth.

▲ Large pieces of Skylab crashed down in Australia in 1979.

▼ When the space shuttle returns to Earth, a large parachute fired from the back slows it down and makes sure it has a happy landing.

◀ The Russian space station *Mir* had an unhappy landing when it burst into flames as it returned to Earth in March 2001.

In April 1967, cosmonaut Vladimir Komarov became the first man to be killed on a space flight. The parachutes on his *Soyuz 1* spacecraft tangled up and did not open. Instead of slowing down, *Soyuz 1* raced towards Earth and crashed at a speed of 145 km/h. Komarov was killed instantly.

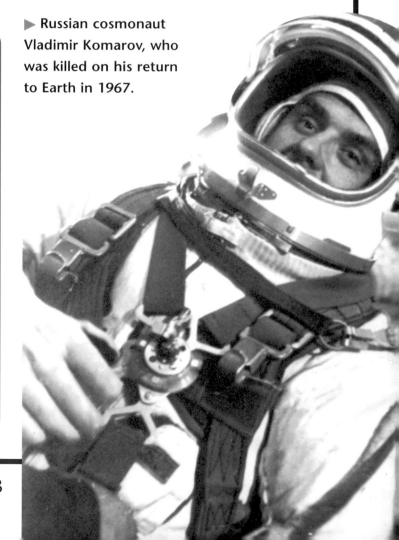

▶ Russian cosmonaut Vladimir Komarov, who was killed on his return to Earth in 1967.

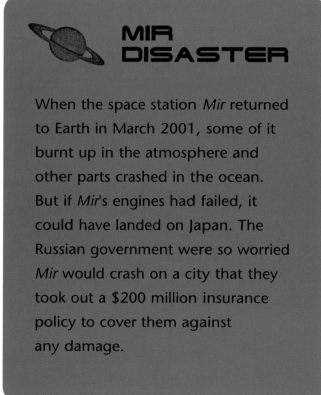

MIR DISASTER

When the space station *Mir* returned to Earth in March 2001, some of it burnt up in the atmosphere and other parts crashed in the ocean. But if *Mir*'s engines had failed, it could have landed on Japan. The Russian government were so worried *Mir* would crash on a city that they took out a $200 million insurance policy to cover them against any damage.

FUTURE DRAMAS

Space travel has always been exciting and dangerous. In the future, astronauts may travel farther and farther from Earth. And who knows what they may encounter – bits of space junk, new galaxies or even aliens!

People have put thousands of objects into space since 1957. Over half of them have since come back to Earth. But quite a lot of 'space junk' such as pieces of old spacecraft still remains in orbit. Space junk moves very fast and it can cause a great deal of damage when it hits something.

▼ The International Space Station has taken years to plan and build. Could it really be destroyed within just a few minutes by space junk?

One danger is that some of the junk could fall back to Earth and kill people. The other danger is that it could stay up in space and crash into satellites or spacecraft. No one on Earth has yet been killed or injured by falling space junk. But because there is so much junk in space the chances of it hitting the new International Space Station are high. Could the next big space drama be caused by space junk?

▲ New discoveries, such as this cartwheel galaxy, show us just how little we know about space and its dramas.

▶ Another future problem for astronauts might be having to fight off aliens!

 MARTIAN MADNESS?

Astronauts who take part in a future trip to Mars may be at great danger from a type of space radiation called cosmic rays. These rays are so powerful that they can damage the human brain and stop a person from ever being able to think properly again. Researchers are now trying to develop computer systems that can check if astronauts are suffering from the effects of cosmic rays.

SPACE-DRAMA FACTS

Blast-off and splashdown dramas

When a *Titan IV* rocket exploded on lift-off in 1998, the noise of the blast set off car alarms over 16 kilometres away. Since *Sputnik 1* took off in 1957, 26 643 objects have gone into orbit around Earth. Some are satellites and spacecraft; some are bits of space junk. Since then 18 000 of these objects have come back to Earth.

A Mercury-Atlas 7 rocket missed its landing spot in May 1962. A rocket failure meant it overshot the landing zone by 400 kilometres.

Computer failure

Two NASA Phobos space probes were lost in 1988. *Phobos 1* moved too far from the Sun, so its solar batteries ran flat. Control of *Phobos 2* was lost when its on-board computer failed.

Sun cycle

The Sun goes through a cycle that comes to a peak every 11 years, with lots of extra sun spots, rays and other activity. This can affect our satellites, radios, mobile phones and electricity supplies here on Earth. The last peak was in 2000.

◄ The *Mars Polar Lander* reached Mars in 1999. Communication was then lost and the mission abandoned.

◀ The space shuttle *Challenger* exploding in 1986.

Death in space

A new Russian rocket was being tested in Tyuratam, Russia, in October 1960. Field Marshal Mitrofan Nedelin arrived to watch the launch. Problems developed before the test, but Nedelin ordered it to carry on. Then the engine caught fire and exploded. The explosion created a poisonous cloud that damaged the lungs of people nearby. At least 91 people were killed.

On 27 January 1967, a fire broke out during a launch test of the *Apollo 1* spacecraft. Three American astronauts, Virgil Grissom, Edward White and Roger Chaffee were killed.

The space shuttle *Challenger* exploded just 73 seconds after its launch. All seven of the astronauts on board were killed. Challenger was one of several space shuttles used by NASA. It had flown nine times before. The disaster was caused by the failure of a seal called an 'O ring' on one of the rocket boosters.

SPACE-DRAMA WORDS

asteroid
A lump of rock or very small planet that orbits (moves around) the Sun.

astronomer
A person who studies the night sky.

blast-off
The launching of a rocket into space (lift-off).

comet
A mixture of ice and rock that orbits the Sun. Comets can be recognized by their long bright tails.

cosmic rays
Dangerous radiation that travels through space.

cosmonaut
A Russian astronaut.

docking
When two spacecraft join together in space.

launchpad
The place where a rocket blasts off into space.

meteorite
Rocks flying around in space.

mission
A voyage to space.

mission control
The job of mission control is to watch over and control a space mission from its base on Earth.

 The long, bright tail of Halley's Comet.

NASA

The National Aeronautics and Space Administration, which organizes space exploration on behalf of the US government.

nuclear reactor

A device that provides power for a spacecraft by using nuclear energy. Nuclear energy is released when certain atoms in the fuel are split apart.

nucleus

The main, leading part of a comet, made of ice and rock.

SAFER jetpack

This stands for Simplified Aid for EVA (extra-vehicular activity) Rescue. It is attached to the base of a normal jetpack. Its jets provide thrust to move the astronaut in different directions.

► The large, square solar panels on this satellite produce electricity to keep it moving through space.

solar panel

A device that changes sunlight into electricity. Solar panels are often used to power spacecraft.

space junk

Broken pieces of rockets and old satellites that stay up in space.

space probe

A spacecraft with no astronauts on board. Space probes are controlled by on-board computers and mission control back on Earth.

space station

A spacecraft that remains in space for a long period of time. Astronauts can live and work in a space station for months or even years.

spacewalk

When astronauts go outside their spacecraft, while in space, to explore space or make repairs.

splashdown

When a spacecraft returns to Earth and lands in the ocean.

umbilical

The safety line that connects an astronaut to a spacecraft during a spacewalk.

SPACE-DRAMA PROJECTS

VISIT A SPACE BASE

NASA has a number of space bases around the United States and some of them are open to the public. If you live in the US or you ever go there for a holiday, perhaps you could take a trip to a space base. You can walk around the base, look at real-life rockets, and find out more about how hard it is to put a rocket into space.

You can find a list of NASA's bases on the Internet at: www.nasa.gov/visitors. html

NASA also tells people about the dates of rocket launches so visitors can watch.

You can find more information at: www.nasa.gov/see_a_launch.html

STUCK ON THE MOON?

Imagine you are an astronaut and you have just taken off from the Moon. You are about to come back to Earth when there is a huge explosion. Your spacecraft seems to be stuck on the Moon! Write a story about your adventure. Can you repair your rocket? How? What items can you find in the rocket to help you? Perhaps you can find some useful spare parts on the Moon? Explain how you get back to Earth.

DRAW A SPACE DRAMA

Pick one of the dramatic stories in this book and try to draw a picture of it. Maybe you could draw the spinning spacecraft, *Gemini 8*, or an astronaut taking a walk out into space. Or you could imagine a futuristic drama of your own and record what happened in pictures.

SPACE DRAMAS ON THE WEB

If you have access to the Internet you will be able to track down information about space including photos of stars, planets, galaxies, spacecraft and information about space missions.

NASA for Kids

www.nasa.gov/kids.html
This site provides information about the stars and planets, rockets, pioneers and astronauts. You can play games and puzzles and take part in on-line activities.

The Space Place

spaceplace.jpl.nasa.gov
A great site with lots of things to make and do and some amazing space facts.

Space kids

www.spacekids.com
An excellent site with photos, videos, facts, games and help with projects for homework.

NOVA Online: Terror in Space

www.pbs.org/wgbh/nova/mir/textindex.html
This site tells you about some of the biggest disasters in the history of space flight.

▲ Tourists watch an astronaut at the Kennedy Space Center in Florida.

NEAR voyage Earth to Eros

near.jhuapl.edu/Voyage/
This site takes you step-by-step on a voyage from Earth to the Eros asteroid.

You can search for other space sites using any search engine. Try using specific search phrases such as the name of a mission, astronaut or spacecraft.

INDEX